捣蛋猫爱数学

中英 双语

什么是形状

Windows, Rings, and Grapes— a Look at Different Shapes

〔美〕布赖恩·P.克利里 ◎著 〔美〕布赖恩·盖布尔 ◎绘 许晓晴 ◎译

形状：
一个东西的外形或者轮廓。

北京师范大学出版社

形状指一个东西的外形，用来形容某个东西的

顶部、

中间

和底部

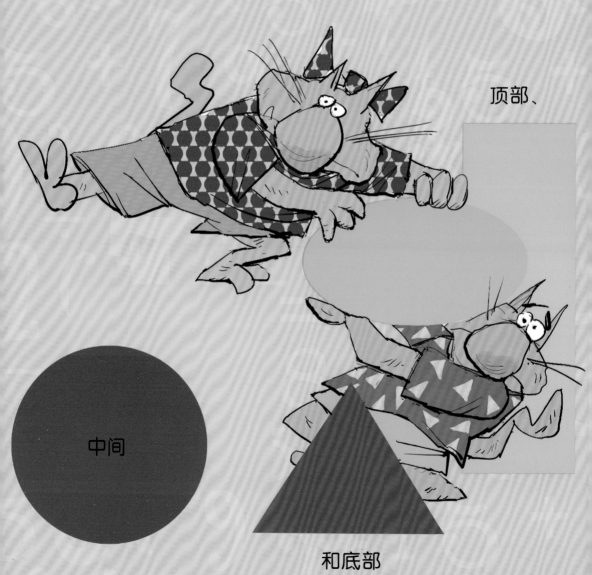

看起来是什么样子的。

里面

外面

彻底弄清楚之后,
你一看到它们,就可以说出它们的形状。

圆形是圆的。

这件长袍上的圆点、

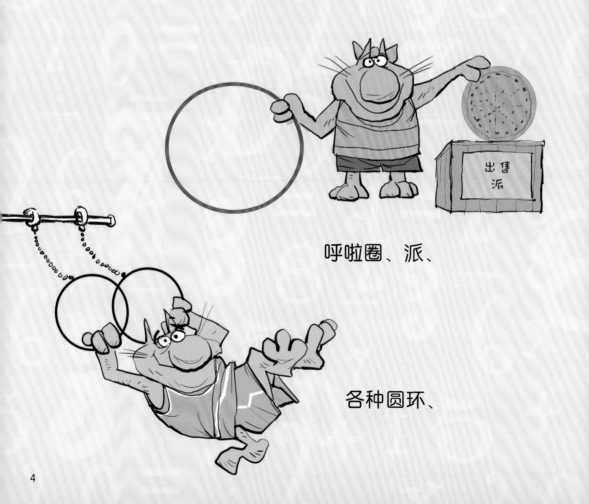

呼啦圈、派、

各种圆环、

烘干机的门

或自行车的轮胎,

所有这些都是**圆形**的。

圆形里没有直线。

它是一条连续的曲线。

它是成圈的、弯曲的。

如果仔细观察，你会发现，

它没有起点，也没有终点。

如果两只小狗想要同一个圆,

它们拉扯这个圆,

这个圆就会变成**椭圆**,

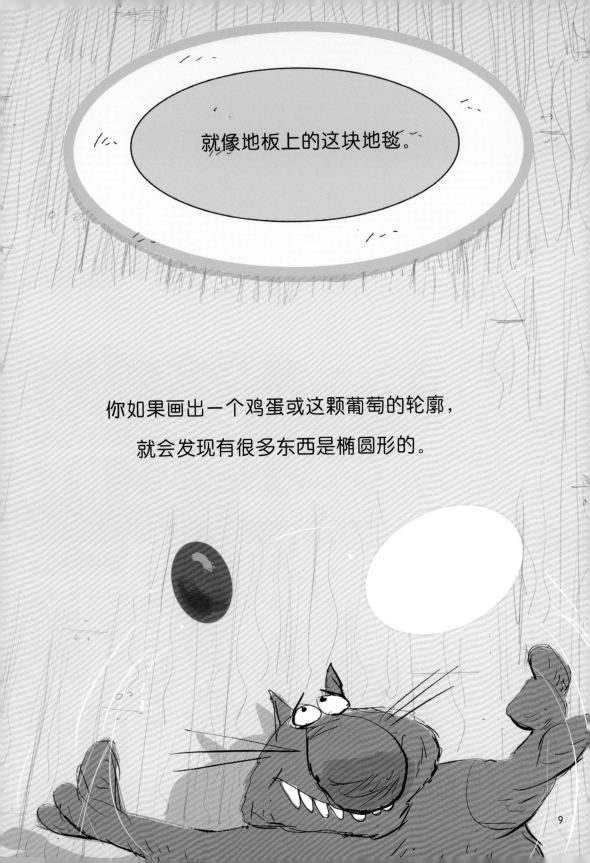

大多数赛道是**椭圆形**的。

如果有一天你去看了，

就会发现，

赛道并不是标准的圆形。

赛道直的部分是比拼速度的好地方，

因为它不是弯曲的，

所以你不必放慢速度。

将三条直线拼接在一起，

结果会很酷哦！

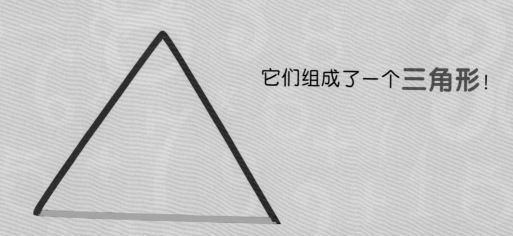

它们组成了一个**三角形**！

他正在加热的这块比萨饼就是**三角形**的。

立在道路和田野相接处
写着"让"字的指示牌呢？

看，它也是**三角形**的。

耳朵上戴的耳饰也是**三角形**的，

每一个都有三条边，

你看出来了吧？

就像三轮车
　　有三个圆形的轮子,
　　　　三角龙
　　　　　　有三个尖尖的角一样,

三角形有三条边,

还有三个角。

试着记住这些形状吧!

你家老房子的窗帘（是什么形状的）呢？

这张地图呢？

这块鼠标垫呢？

它们都是**正方形**的。

这个形状由四条边组成。

你会发现，

它还能分成四个小**正方形**。

正方形所有的角看起来

都和书的角一样——

有点儿像大写的 L。

你看,
就像这家新酒店的窗户一样,
正方形所有的边都一样长。

如果其中两条边短一些，
另外两条边长一些，

那它就是**长方形**。

这扇门、

这张来自挪威的明信片、

这个屏幕、

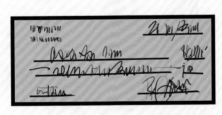

这张支票以及这个钱包,都是**长方形**的。

看，这块苹果汁广告牌也是**长方形**的，
它跟**正方形**有点儿像，
只是更宽一些。

这面墙和这只类人猿拿着的画框也是这样，

有点儿像**正方形**，

只是更高一些。

所以，

当你看到地毯、

窗帘、

午餐盒、

背包或者

游戏上的图形时，

你就知道它们叫什么——

不管是大的还是小的——

因为你会记住这些图形的名称!

那么，什么是**三角形**？

什么是**长方形**？

什么是**正方形**？

什么是**圆形**？

什么是**椭圆形**？

你都知道了吗？

三角形有三条边和三个角。

长方形有四条边和四个角，
每个角都是直角。

正方形有四条一样长的边
和四个角，每个角都是直角。

圆形是一个完美的圆，
没有角。

椭圆形就像一个鸡蛋，
没有角。

Windows, Rings, and Grapes—a Look at Different Shapes

第2~3页

A shape is a form. It's how something looks, with a top and a middle and bottom. And knowing about them all inside and out, they're easy to name when you spot 'em.

第4~5页

A circle is round, like the dot on this gown, like a hula hoop, pie, or some rings, the door on this dryer, a bicycle tire. The circle forms all of these things.

第6~7页

There's not one straight line in a circle's design. It's just a continuous bend. It's looped and it's curved, and when closely observed, you'll find no beginning or end.

第8~9页

If two dogs would fetch one and lengthen and stretch one, an oval would be its new shape. Like this rug on the floor, and you could make more if you outlined an egg or this grape.

第10~11页

Most racetracks are oval. If one day you go, you'll note stretches that aren't quite as round. These straightaway places are perfect for races, with no curves to make you slow down.

第12~13页

When three straight lines meet, the result is quite neat. A triangle shape will be formed. This nice slice of pizza he's trying to heat's a triangular treat being warmed.

第14~15页

That sign that says "Yield" where the road meets that field? See, that's a triangle as well. These earrings that dangle are each a triangle, 'cause both have 3 sides, you can tell!

第16~17页

So, just as a tricycle has 3 round wheels and triceratops has 3 sharp horns, triangles have both 3 sides and 3 corners, so try to remember these forms!

第18~19页

The shade your old house had? This map? ... or this mouse pad? Each one of these shapes is a square. The form is four-sided and when it's divided, you'll find four more squares inside there.

第20~21页

The corners all look like the edge of this book—a bit like an uppercase L. The sides have to be all the same length, you see, like the windows in this new hotel.

第22~23页

If two sides were smaller and the other two taller, a rectangle is what you'd call it. It's the shape of this doorway, this postcard from Norway, this screen, and a check and this wallet.

第24~25页

Like a square, only wider, this billboard for cider is rectangular in its shape. Like a square, only taller, so is this wall or the frame for this fine painted ape.

第26~27页

So, when you see shapes on the carpet or drapes, on your lunch box or backpack or games, you'll know what to call every one—big or small—because you'll know each of their names!

第28~29页

So, what is a triangle? What is a rectangle? What is a square? What is a circle? What is an oval? Do you know?

A triangle has three sides and three corners. A rectangle has four sides and four corners. The corners are right angles. A square has four equal sides and four corners. The corners are right angles. A circle is perfectly round. It has no corners. An oval is a shape like an egg. It has no corners.

献给埃伦。
——布赖恩·P.克利里

Text copyright © 2009 by Brian P. Cleary
Illustrations copyright © 2009 by Lerner Publishing Group, Inc.
Simplified Chinese rights arranged through CA-LINK International LLC (www.ca-link.com)
English and Chinese bilingual translation © 2019 by Beijing Science and Technology Publishing Co., Ltd.

著作权合同登记号　图字：01-2018-2347

图书在版编目(CIP)数据

捣蛋猫爱数学.什么是形状/（美）布赖恩·P.克利里著；（美）布赖恩·盖布尔绘；许晓晴译. —北京：北京科学技术出版社，2019.3
ISBN 978-7-5714-0031-6

Ⅰ.①捣… Ⅱ.①布… ②布… ③许… Ⅲ.①数学-儿童读物 Ⅳ.①O1-49

中国版本图书馆CIP数据核字（2019）第002055号

捣蛋猫爱数学.什么是形状

作　　者：〔美〕布赖恩·P.克利里	绘　者：〔美〕布赖恩·盖布尔
译　　者：许晓晴	策划编辑：石　婧
责任编辑：樊川燕	责任印制：张　良
出版人：曾庆宇	出版发行：北京科学技术出版社
社　　址：北京西直门南大街16号	邮政编码：100035
电话传真：0086-10-66135495（总编室） 　　　　　0086-10-66161952（发行部传真）	0086-10-66113227（发行部）
电子信箱：bjkj@bjkjpress.com	网　址：www.bkydw.cn
经　　销：新华书店	印　刷：北京宝隆世纪印刷有限公司
开　　本：710mm×1000mm　1/16	印　张：2
版　　次：2019年3月第1版	印　次：2019年3月第1次印刷
ISBN 978-7-5714-0031-6/O·032	

定价：25.00元

京科版图书，版权所有，侵权必究。
京科版图书，印装差错，负责退换。